black dog

THE CHIMPANZEE BOOK

APES LIKE US

by Dr Carla Litchfield

Dr Carla Litchfield is a scientist at ZoosSA, and lecturer at UniSA. She has studied the great apes in the wild in Uganda and in zoos and sanctuaries around the world. She is the current President of the Australasian Primate Society.

For my wonderful daughter, Kaitie Afrika Litchfield—my inspiration, my love, my life. Thanks to the Kanyawara community of chimpanzees at Kibale Forest for sharing their world with me for a year.

Photo credits:
Michael Nichols/National Geographic: front cover, pp i, 2, 14, 29, back cover;
Big Stock Photo: pp ii, 6;
Kenneth Garrett/ National Geographic: pp iii;
Cyril Ruoso/Minden Pictures/National Geographic: pp 2, 8, 9, 12, 17, 28;
shutterstock pp 3, 11, 21, 22, 26;
DLILL/Corbis: p4;
Carla Litchfield: pp 6, 25;
Istock: p7,
Gerry Ellis/Minden Pictures: pp 6, 14, 15;
Frans Lanting: pp 8, 16, 20, 21, 24;
Stan Osolinski/Photolibrary: p13;
Etsuko Nagami: p19;
Dr Kimberly J Hockings: p18-19,
aapimages: pp 22, 23, 29, 30;
Michael Poliza/ National Geographic: p27;
Ian Redmond: p27;

First published in 2009 by
black dog books
15 Gertrude Street
Fitzroy Vic 3065
Australia
61 + 3 + 9419 9406
61 + 3 + 9419 1214 (fax)
dog@bdb.com.au

Designed by Blue Boat Design
Printed and bound in China by Everbest Printing

FSC is a non-profit international organisation established to promote the responsible management of the world's forests.

National Library of Australia
cataloguing-in-publication data:
Litchfield, Carla.
The chimpanzee book: apes like us.

Includes index
For primary school children.
Subjects: Chimpanzees--Juvenile literature
ISBN: 9781742030746 (pbk)
Dewey number: 599.885

10 9 8 7 6 5 4 3 2 1 9/0 1 2

KEY

On some pages you will find a sidebar which shows: how far above sea level chimps live; the names of some other **primates** in the same area, and chimpanzee predators. This information is always changing due to war, disease, loss of habitat and poaching.

black dog books would like to thank Professor Colin Groves for his thorough factual check of this book.

CONTENTS

OUR CLOSEST RELATIVES

Chimps are our closest animal relatives. They belong to the great ape family, along with humans, gorillas and orangutans. There are two **species** of chimps, called chimpanzees (*Pan troglodytes*) and bonobos (*Pan paniscus*). **Genetically**, chimps are more closely related to humans than they are to gorillas.

In this book we will use the word 'chimps' to refer to chimpanzees and sometimes to bonobos as well.

Like humans, chimps have large and complex brains. They have the ability to use 'language', to make and use tools and to recognise themselves in mirrors. They have traditions that are passed from one generation to the next. These abilities were all once thought to be unique to humans.

Like us, they kiss, hug and comfort each other. They have a very long childhood, form lifelong friendships and grieve for their dead. They can also hold a grudge for years, deliberately hide their feelings and even go to 'war' against their 'enemies'.

Chimpanzees and bonobos are **omnivores**. They eat plants, ants and termites, and also hunt monkeys and other small mammals. Bonobos do not seem to hunt as much as chimpanzees.

A squirrel monkey

Chimpanzees and bonobos are apes, not monkeys. Apes do not have tails, and can swing by their arms between tree branches. The shoulder bones of most monkeys don't allow them to swing the same way.

WHERE IN THE WORLD?

Chimps are great at finding their way around. They have 'mental maps' so that they can remember the location of different fruit trees, streams, paths, caves, termite mounds, and other important parts of their environment.

Chimpanzees and bonobos are just as at home on the ground as they are in the trees. Bonobos tend to spend more time in trees.

Chimpanzees live in rainforests and **savannah**-woodlands of west, central and east Africa. One hundred years ago there were more than a million chimpanzees living across Africa. Now there are no more than 300 000. Bonobos are only found in the Central Basin rainforest of the Democratic Republic of Congo. There are about 10 000 bonobos left.

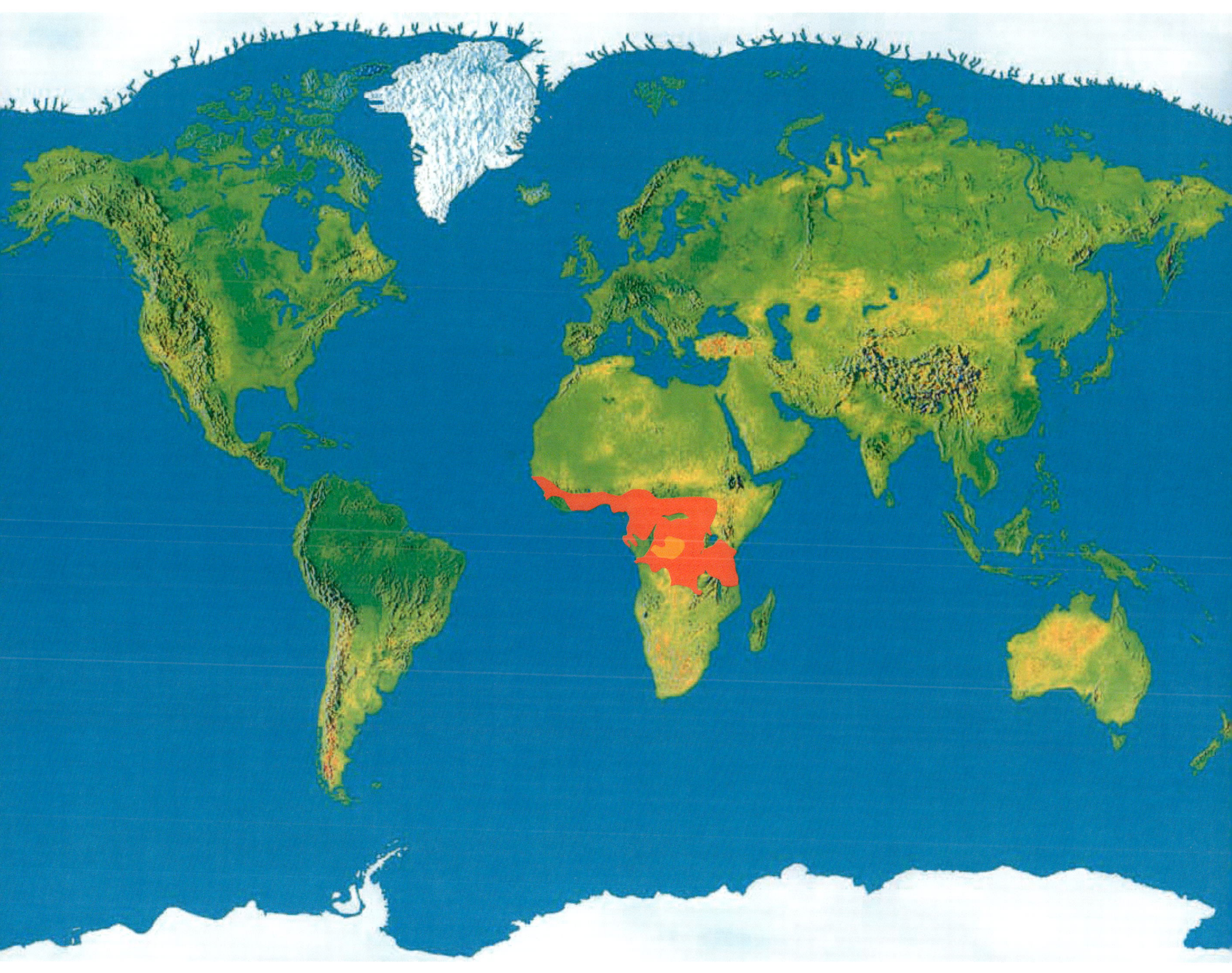

Chimps are found across these countries, in remaining patches of **habitat**. In some countries, there are only tiny groups of chimps left.

Where bonobos live

Where chimpanzees live

CHIMPANZEE BITS

Adult males are a bit bigger than adult females. Males weigh between 34–70kg. They can be taller than 1 metre.

They have **opposable** thumbs and big toes, which means they can grip things with both their hands and feet.

Their arms are longer than their legs, so their back slopes down when they walk on their knuckles.

Youngsters can be identified by a white tuft of hair on their bottoms. This signals to older chimpanzees to be gentle, and allows the young chimpanzees to get away with being naughty.

Adult females weigh between 26–50 kg. They are less than 1 metre tall.

Babies have pink faces, which darken as they get older.

Adult chimpanzees are about five times stronger than adult humans.

BONOBO BITS

Their 'hairstyle' is long and parted in the centre of their head.

Babies have black faces just like adults. They also have pink lips.

They have hardly any hair on their chins, but have cheek whiskers that stick out.

Their feet are sometimes webbed between the second and third toes.

Bonobos are more slender and graceful-looking than chimpanzees.

Bonobos usually weigh less and have slightly smaller heads than chimpanzees.

Their legs are long, so their backs are flat when they walk on the knuckles of their hands.

BEHAVIOUR, NOISES & FACIAL EXPRESSIONS

Chimpanzees and bonobos communicate with body language, facial expressions, and sounds. Young chimps playing or being tickled laugh noisily with open mouths. Grumpy chimps purse their lips together and glare, and if they are really angry their hair stands on end, making them look fearsome. They stamp their feet and throw things around when they have a temper tantrum. When adult males fight, they bite, kick, thump and punch.

Chimps whimper and scream when scared or upset. They make loud 'pant-hoots' to let others know where they are. Young chimps often make 'hoo' sounds to stay in touch with their mothers. When enjoying a snack, chimps make food-grunts—a bit like we do when munching a treat.

Chimpanzees never smile by showing all their teeth and gums. A wide chimpanzee 'smile' is actually a fear grimace.

Just like humans, every chimp is unique. You can tell them apart by their faces, bodies, voices, and the way they move. Their hair varies in colour, with different shades of brown or black, and they go grey and bald at different ages.

Some have the muscled, powerful limbs of a bodybuilder, while others are quite thin. Some chimps are shy and scared of new things. Others are 'thrill seekers', rushing to investigate anything new. Some are patient and gentle, while others are bullies and easily angered.

GROWING UP IN THE FOREST

Chimps eat mostly fruit. When it is scarce they split into small groups and travel in search of food. When fruit is plentiful they come together in large groups. The area erupts with loud pant-hoots, screams and buttress-drumming (made by hitting trees with hands and feet).

Older sisters and brothers help their mums look after new babies. Chimps spend at least eight years close to their mothers.

When female chimps grow up, many leave home and join another community. Males stay in their home community. Chimps usually live in groups of about 50, but can live in communities of more than 100. Some chimp communities can have as few as 10 individuals.

MAKING AND KEEPING FRIENDS

Playing is an important part of growing up to be a healthy and happy chimp. Just like us, they play different games—they spin around until they are dizzy, they love being tickled, they chase each other, and they drag sticks around. Some youngsters even carry log 'dolls', looking after them and treating them like babies.

For male chimpanzees, and bonobos of both sexes, friendships are vital in securing a top spot in the community.

Chimps spend a lot of time working on friendships that can last a lifetime. Grooming another chimp shows that you like him or her. If you spend enough time grooming another, then you can be counted on for support, and the chimp you are grooming is more likely to support you.

CHIMPANZEE CULTURE

Scientists in Africa have been observing chimps at more than 10 different sites, and have found that each chimp community has its own behaviours and traditions. Chimpanzees make and use tools, but the tools they use vary between groups. Some use sticks or grass stems to collect termites and ants. Leaves are used as sponges to clean themselves, or are '**clipped**' to get another chimpanzee's attention. Unfortunately, we know a lot less about bonobos because they live in areas where it is too dangerous for scientists to stay for long periods.

Some chimps use stones as a **'hammer and anvil'** to crack open nuts.

Some chimps hate water; others love it. Before or during storms, some even perform a rain dance!

A chimpanzee uses a coconut shell to scoop up water.

Scientists discovered that chimps bring special tools to the termite mounds—short sticks to dig into mounds, and more delicate 'fishing probes' to get the termites out.

Studying wild chimps is difficult. Chimps live in remote areas, are hard to see, and can take months or years to **habituate** to humans. Thanks to new technology, scientists have set up camera traps to watch wild chimpanzees and bonobos without disturbing them.

At the Goualougo Triangle in the Congo, cameras were set-up at termite mounds. Some chimpanzees were scared of the cameras, but others inspected them, even poking sticks at the lenses. Scientists have radio-collared **rehabilitated** orphaned chimps to track how they behave when released back into the wild.

MOST FAMOUS APES

GOMBE CHIMPANZEES

Dr Jane Goodall began studying wild chimpanzees at Gombe National Park in 1960. At this time, because she was a woman, Dr Goodall was only allowed to go to Gombe if she had someone else with her. So for the first four months her mum, Vanne, went with her.

Dr Goodall was the first scientist to report that chimpanzees make and use tools, when she observed a chimpanzee she named David Greybeard stripping leaves from a twig before fishing for termites.

Where:	Tanzania
Map Coordinates:	4°40′ S 29°38′ E
Rainfall:	1600mm per year

Over the years, Gombe chimps were observed to wage 'war' against other chimpanzee communities. Over a four-year period, the 'Kahama' group of chimpanzees were systematically killed by the 'Kasakela' group.

Flo and Fifi's family, Gombe

Dr Goodall gave the chimpanzees names, and recognised them as individuals—not just as numbers or 'things' as many scientists did at the time. As a result, people all around the world have been able to follow the stories of these chimps, such as Flo and her daughter Fifi's family, the 'F' family, for more than 40 years.

Fifi's son Frodo turned out to be a bully and an accurate stone-thrower. Flo was an amazing mother, and when she died of old age her eight-year old son Flint became depressed and died, seemingly of a 'broken-heart'.

Dr Goodall helped people see that, like humans, chimpanzees could feel emotions and pain.

600m

- Habitat: Thick riverine forest, woodland and open grassland areas
- Predators: Leopards and humans

400m

- Other primate neighbours: Bushbaby, blue monkey, red-tailed monkey, olive baboon, vervet monkey, red colobus monkey
- First studied by: Dr Jane Goodall

200m

Height above sea level

PEACEFUL APES
BONOBOS

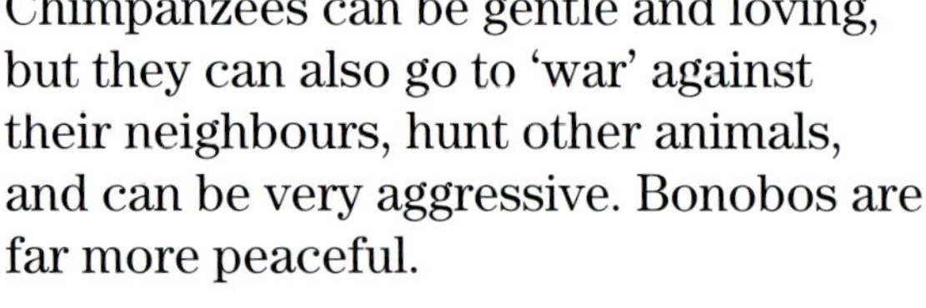

Chimpanzees can be gentle and loving, but they can also go to 'war' against their neighbours, hunt other animals, and can be very aggressive. Bonobos are far more peaceful.

Where:	Democratic Republic of Congo
Map Coordinates:	0°11′ N 22°28′ E (Wamba) & 0°51′ N 21°5′ E (Lomako)
Rainfall:	2000mm per year (Wamba), 1850mm per year (Lomako)

When neighbouring bonobo communities meet, the encounters are friendly and not war-like.

Bonobos are very affectionate, using hugs, cuddles and kisses to resolve tension rather than fighting.

Bonobos have equal power between adult males and females. To become leader, a male bonobo needs the help of his mother and her female friends. Whenever a new female moves into the community, she makes friends with the other females first.

Bonobos spend more time in the trees than most chimpanzees. They rarely hunt, and seem to spend more time walking upright like humans. Bonobos have squeaky bird-like voices and are more "chatty" than chimpanzees.

800m

Habitat:	Mosaic of rainforest, swamp, cultivated fields (Wamba); evergreen rainforest and swamp forest (Lomako)
Predators:	Leopards, humans

600m

400m

Other primate neighbours:	Bushbaby, potto, red-tailed monkey, Wolf's mona monkey, Salongo monkey

200m

First studied by:	Dr Takayoshi Kano, Dr Noel Badrian, Dr Alison Badrian

Height above sea level

CROSSING ROADS SAFELY

BOSSOU CHIMPANZEES

Around 2500 people live in the village of Bossou, and a group of about 12 chimpanzees lives in the surrounding forest. When moving between forest patches searching for food, the chimpanzees have to cross two roads. One of the roads is narrow and used by pedestrians. The other bigger road is used by trucks, cars and motorbikes.

Just like humans, the chimpanzees have learned to stop, look, and wait for a gap in the traffic. The three adult males travel at the front and back of the group, protecting the females and youngsters.

Where:	Guinea
Map Coordinates:	7°39′ N 8°30′ W
Rainfall:	2000–3000mm per year

This photo shows Yolo, the **alpha male**, who is usually at the back of the group. Often Foaf or Tua, the other adult males, stay on the road and check for traffic until the whole group is across safely—just like lollipop people at a school crossing!

Scientists observed and timed the chimpanzees waiting for about half a minute before crossing the small road safely. The chimpanzees understand that the big road is more dangerous, and wait at least three minutes before crossing.

800m

600m

400m

200m

Habitat:	Patchy mosaic of forest, savannah and cultivated fields
Predators:	Humans
Other primate neighbours:	Unknown, mainly humans
First studied by:	Dr Yukimaru Sugiyama. Dr Kimberley Hockings, Dr James Anderson and Dr Tetsuro Matsuzawa conducted the 'crossing roads' study

Height above sea level

NEW KIDS

FONGOLI CHIMPANZEES

The scientific world was turned on its head with reports of chimpanzees that lie around in water to cool down, shelter in caves to escape the heat and seem to make spears to hunt.

Fongoli chimpanzees live in a hot, dry savannah-woodland environment, very like what early humans experienced. Trees are short and spread out, and during the day temperatures can reach 42°C.

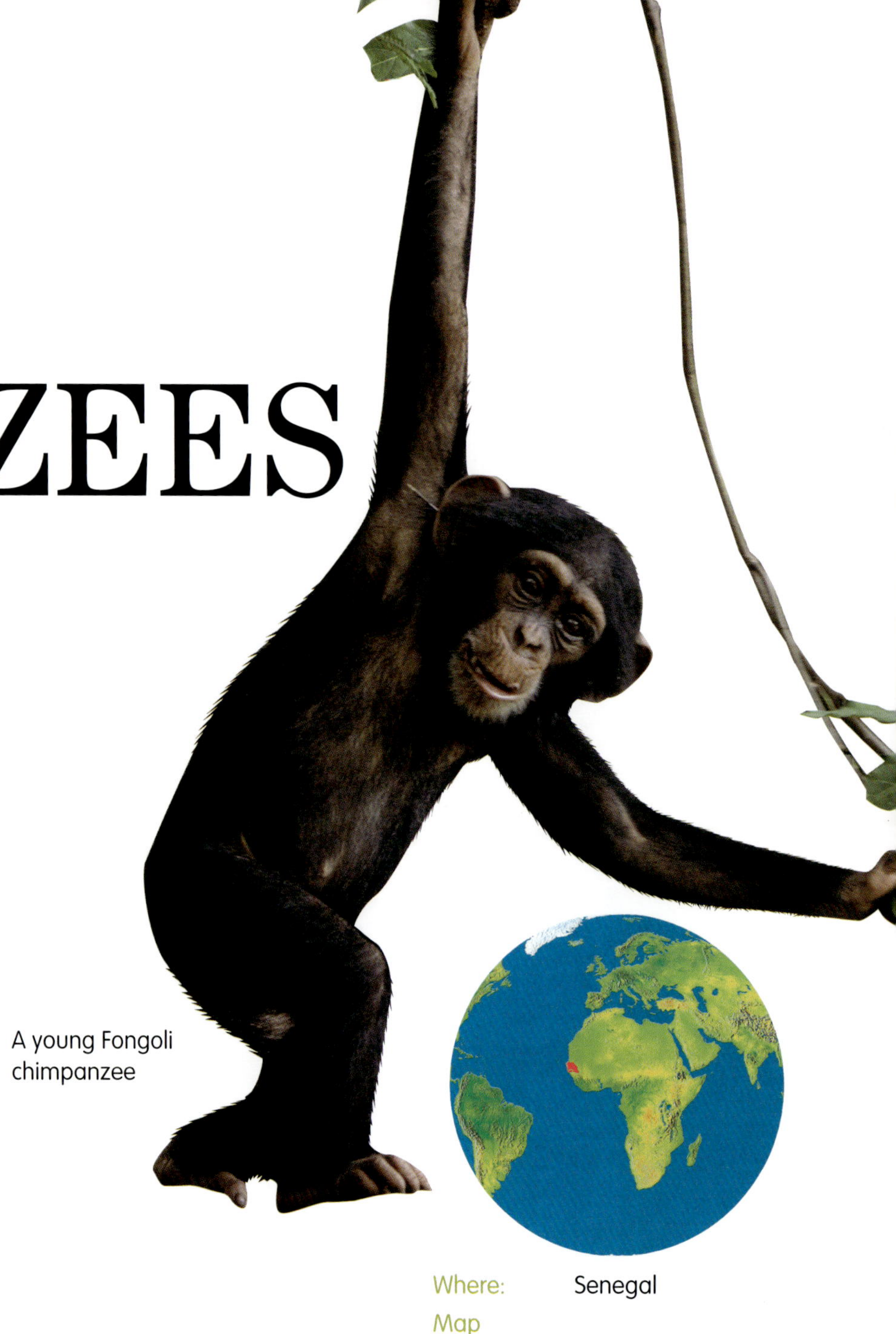

A young Fongoli chimpanzee

Where: Senegal

Map Coordinates: 12°40′ N 12°12′ W

Rainfall: 900–1100mm per year

A bushbaby

Females and young chimpanzees stab at tiny bushbabies with sharpened sticks, a bit like spears. Bushbabies are **nocturnal**, sleeping in tree hollows during the day. When chimpanzees find a tree hollow, they break off a branch, strip the leaves, and make a pointy end with their teeth, before jabbing the tool into the hollow.

This stunned the scientific world, because Fongoli chimpanzees were actually making a tool or weapon to hunt. Such tool making was once thought to be unique to humans. Chimpanzees make us really think about what it means to be human.

800m

Habitat: Mosaic of open grassland savannah, woodland, gallery forest

Predators: Humans

600m

400m

Other primate neighbours: Bushbaby, humans

First studied by: Dr Jill Pruetz, Dr Linda Marchant, Dr Bill McGrew, Dr J. Arno

200m

Height above sea level

FIGHTING OFF LEOPARDS
TAÏ CHIMPANZEES

Chimpanzees at Taï have strategies to prevent being attacked by predators. If chimpanzees hear a leopard or the screams of a friend being attacked, they rush to chase the leopard away. If chimpanzees get hurt by a leopard, their friends clean the wound and take care of them.

Sometimes chimpanzees will use broken branches as weapons against leopards.

Where: Ivory Coast

Map Coordinates: 15°52′ N 7°20′ W

Rainfall: 900–1100mm per year

In the 1960s, Dr Adrian Kortlandt observed wild chimpanzees in Uganda responding to a stuffed leopard with a baby chimp doll in its paws. These photos show an angry chimpanzee mother throwing sticks at the fake leopard.

800m

Habitat: Evergreen moist rainforest

Predators: Leopards, lions, humans

600m

Other primate neighbours: Bushbaby, lesser white-nosed monkey, Diana monkey, greater spot-nosed monkey, sooty mangabey monkey, bay colobus monkey

400m

First studied by: Dr Christophe Boesch and Dr Hedwige Boesch

200m

Height above sea level

LEGENDARY GIANTS?

BILI CHIMPANZEES

A plaster cast of a big ape footprint.

People love to think that somewhere in the world there is still a giant ape, a 'Big Foot' or 'Yeti', to be discovered. In the late 1990s, rumours emerged of a new ape species, a giant gorilla-like chimpanzee.

Locals near the Bili Forest told tales of large chimps that 'kill the lion'. Evidence of ground nests, large footprints and skulls, and a couple of photos suggested that these were large chimps that behaved like gorillas. **DNA** tests conducted on hair and poo samples confirm that the 'mystery apes' are chimpanzees. Ongoing studies will hopefully find out more about them.

A 'trap shot' or hidden camera photograph of a 'mystery' ape.

Where: Democratic Republic of Congo

Map Coordinates: 4°9′ N 25°10′ E

Rainfall: 1200+mm per year

800m

600m

Habitat: Mosaic savannah, woodland, tropical rainforest and cultivated areas

Predators: Leopards, lions, humans

400m

Other primate neighbours: De Brazza's monkey, olive baboon, crowned guenon monkey, grey-cheeked mangabey monkey

200m

First studied by: Karl Ammann (photographer) Thurston Cleveland Hicks

Height above sea level

SMART CHIMPS: PROBLEM SOLVING, THINKING AND COMMUNICATING

Kanzi and Dr Sue Savage-Rumbaugh

We know a lot about chimps by studying them in captivity. Between 1913 and 1917, Dr Wolfgang Köhler studied how chimpanzees solved problems. The chimpanzees had to grab a banana hanging just out of their reach. After some unsuccessful attempts, they reached the banana by stacking boxes or using poles. They showed evidence of planning, thinking and skilful problem-solving.

Kanzi is a famous bonobo who has a **vocabulary** of about 500 words. He understands spoken English like a two or three-year-old human child, and communicates using symbols on a keyboard. He can even make Stone-Age tools. Once, on a trip to a forest with his teacher Dr Sue Savage-Rumbaugh, he made a fire using sticks and matches and then toasted his own marshmallows!

Just like us, chimps get headaches, toothaches, stomach-aches, cuts and bruises. They can't go to a doctor for help, but they do know which herbs and plants to use as medicine. Many plants that chimps use have been analysed for their chemical properties, leading to the discovery of new drugs than can help humans.

Ai and her baby Ayumu

Young chimps have the same problems solving the 'trap-tube' task as three or four-year-old human children. They have to put the tool in the opposite side to a reward to push it out, or it will fall into the trap where they can't get it. This chimp is about to get it wrong!

Since 1978, a chimpanzee called Ai has been using a computer to communicate with scientists in Japan. Ai learnt all 26 letters of the English alphabet, and more than 40 Japanese Kanji characters. She uses more than 80 visual symbols as 'words', and numbers from zero to nine. Like his mother Ai, Ayumu is skilled at using a computer to communicate with scientists.

CHIMPS UNDER THREAT

There is still much to discover about chimpanzees and bonobos, but sadly they are disappearing before we get the chance. Chimpanzees have already vanished from four African countries.

One problem for all apes is the destruction of their **habitat**. Some people outside Africa want to buy tropical wood, rather than using plantation timbers. Another problem is the illegal mining for minerals such as **coltan**, which is used in computers and mobile phones.

An estimated 4000 chimpanzees are killed every year for 'bushmeat', an illegal delicacy for wealthy people living in big cities in Africa, Europe, the UK and USA.

PASA sanctuaries care for orphaned gorillas and chimps, who may be released back into protected forests.

When a logging or mining company moves into a rainforest, roads may be built into areas where humans previously did not go. Camps are set up for workers, who also need to be fed. So hunters are employed, who gradually kill all the animals living nearby. This creates opportunties for hunters to make money by sending dead animals, or 'bushmeat', to the main cities.

Baby chimps are taken screaming from their dead mothers to be kept or sold as pets. These traumatised orphans usually die. Sometimes, if they are lucky, they are confiscated by authorities and taken to a **PASA** sanctuary.

Chimps make terrible pets. They are cute when they are babies, but as toddlers they can destroy a house. By four or five years old they are stronger than human adults, and can bite, hit and throw tantrums.

SAVING THEM FROM EXTINCTION

Young human children are usually not allowed to visit wild chimpanzees and bonobos, as young chimps can catch childhood diseases like chickenpox, mumps, measles and the flu.

The demand for coltan is high, since many people get a new mobile phone every year. An easy way to help stop the mining of coltan is to encourage recycling of old mobile phones.

Buying **fair trade** African products, such as chocolate, tea or coffee, can also help. 'Gombe Special Reserve' coffee is 'chimp-friendly'. It is grown by farmers on tiny blocks of land near the Gombe chimpanzee site. The Jane Goodall Institute found a way of paying coffee growers a fair price, so that they could help protect their environment and still make a living.

Thousands of young people are helping animals and the environment through the Roots and Shoots program. This global network started in 1991 with 16 Tanzanian teenagers who, along with Dr Jane Goodall, wanted to help make their world a better place. You can join or start a group, and choose a project to support. You can also share ideas and get in touch with other groups around the world.

Zakayo celebrates his 44th birthday with some cake at the Uganda Wildlife Education Center.

If you are lucky enough to visit Africa, find out about tourist sites where you can see chimps. In Uganda, tourists pay a lot of money to visit chimps in the wild or orphaned chimps in sanctuaries. This money helps support local communities and shows that people all around the world care about chimpanzees and bonobos.

Tourists should not buy anything made of animal products, and must never buy an orphaned chimpanzee or bonobo—to do so will mean hunters will just try to make more money by taking more chimp babies.

All of us can help save chimpanzees and bonobos if we are careful about what we buy and how we treat our environment. Finding out more about sanctuaries or raising money for orphaned chimps can help.

GLOSSARY AND INDEX

Further Resources

Chimp sounds, download activities and Gombe tour:
http://www.discoverchimpanzees.org/activities/activities.php

Find out more about Dr Jane Goodall and the Roots and Shoots program:
http://www.janegoodall.org/
http://www.rootsandshoots.org/

To see what Kanzi is up to, and to find out more about Great Apes, visit the Great Ape Trust:
http://www.iowagreatapes.org/index.php

Ai's homepage:
http://www.pri.kyoto-u.ac.jp/ai/index-E.htm

For updates about Bossou and Nimba chimps, and the Green Corridor Project:
http://www.greenpassage.org/indexE.rhtml

Information, newsletters and activities about chimpanzees of Taï Forest:
http://www.wildchimps.org/wcf/english/pan/index.html

Find out about PASA sanctuaries:
http://www.pasaprimates.org

Glossary

alpha-male: the 'top' male or leader of a community.

anvil: usually a rock or flat surface that a nut sits on while the chimp bashes it with the '**hammer**' (usually another rock).

clipping (of leaves): Chimps use their mouth or fingers to noisily rip small pieces off a leaf. They do this when stressed or frustrated, or to attract another chimp's attention.

coltan: columbite-tantalite, a metalic ore used in many electronic products.

DNA: deoxyribonucleic acid, which contains all the genetic information about each living organism—what it is and what it looks like.

fair trade: payment of a fair price for goods, especially for products from developing countries. Farmers can then make a living by growing crops, and so don't have to hunt illegally or cut down trees.

genetics: the study of the variation of genes, which are passed from parents to children and may influence how we look and behave.

habitat: the natural environment of an animal.

habituate: to remove fear of humans. If chimps keep seeing people who don't hurt or disturb them, they will gradually get used to having people around.

mammal: warm-blooded animals with backbones. Most give birth to live young rather than laying eggs. They suckle their young with milk.

nocturnal: mainly active at night and asleep during the day.

omnivore: eats plants and animals.

opposable: the tip of an opposable big toe or thumb can touch the tips of other toes or fingers, allowing the ape to grip things and pick them up.

PASA: Pan African Sanctuary Alliance

predator: an animal that hunts or feeds on another animal (which is the 'prey').

primate: a group of mammals which includes great apes, humans and monkeys.

rehabilitate: some orphaned apes can be returned to the wild after they are healthy. They need to be able to find food, climb trees, build nests, and show other survival skills.

savannah: grassland region with scattered trees.

species: a biological classification for a group of organisms that interbreed and produce fertile babies.

vocabulary: a set of words that are understood or used as 'language'.

Index